CONVERT YOUR CAR TO ALCOHOL

By Keat B. Drane

Marathon International Book Company
P.O. Box 40
Madison, Indiana 47250-0040 U.S.A.
Telephone (812) 273-4672
Fax (812) 273-8964
www.MarathonBooks.Biz

Library of Congress Card Number: 80-81750

ISBN –13: 978-0-915216-61-1
ISBN--10: 0-915216-61-2

Disclaimer
These modifications and costs for
conversion are factual and based on the prices in 1979.
The author converted a 1969 Dodge Dart
(owned by the author's friend, Jim Wortham) to run on
pure alcohol, with no gasoline added. The author and
publisher request that you **do not use these modifications
on fuel-injected and computerized vehicles. Also, please
do not use these modifications on vehicles with
diesel-fuel engines.**

For an update to this book, please order:
Alternative Fuel Update #1
Price $5.00 postpaid.
This report provides updated resources and information
on the use of alcohol for modern vehicles, how vegetable oil
may be used for diesel-fuel vehicles and information on
how to convert a car to all-electric (battery power).
You may order this report from the address on top of this page.

CONTENTS

Figure 1-1—This is the test car, a 1969 Dodge Dart.

ALTERNATIVE FUEL SOURCE

Are you as tired of the "gasoline shortage" as I am? Long lines at the pump, shorter service station hours and higher prices are here to stay. Rationing has affected various areas of the nation. Unstable, international politics threaten our continued dependence on foreign crude oil. Politicians and oil company executives predict the situation to worsen.

Faced with these grim facts, I began to search for an alternative fuel source. I wasn't looking for a solution to the entire country's energy problems. I simply wanted to insure that my family and I would have the fuel to power our vehicles when and where necessary, without interference or restriction from Uncle Sam's energy "experts" or the whims of a greedy oil sheik in the Middle East.

Alcohol fuel is the little man's best hope for relief from gas "pains". This all-but-untapped, domestic resource has many advantages. The following is a list of the ones I feel are most significant:

1. Almost any gasoline-powered engine can be made to run well on alcohol.
2. Only minor and inexpensive modifications to the engine are required.
3. Anyone with reasonable, mechanical skill and common handtools can make the modifications once they've learned the procedure.
4. Alcohol can be produced from a variety of organic materials and is a natural substance.

10

5. Distilling can be done on a small scale by individuals or on a very large scale by local companies.
6. Any profit generated by production of domestic alcohol fuels will stay in America and will pay American laborers.

With these thoughts in mind, I decided to undertake a test project. The first goal was to successfully convert a vehicle to alcohol fuel rather than "gasohol". The vehicle was then to be driven daily to determine if alcohol was practical for use under normal everyday conditions.

THE TEST VEHICLE

The test vehicle was a 1969 Dodge Dart. It had 80,000 miles on the motor at the time of conversion and was stock in every aspect. The Dart was equipped with a 318 cubic inch engine, automatic transmission, power brakes and steering and air conditioning.

The Dart was donated for testing purposes by friends who shared my interest in finding a practical alternative to gasoline.

The conversion and test were a great success. I now feel secure in the knowledge that I'll never have to park my car for lack of fuel. I want to share my good fortune and results with you. The following pages outline the theory, mechanics, cost and materials which you should follow in converting your car.

Let me give you a word of caution. Please read this entire book before starting to modify your car. There are several variations discussed, and you will want to be sure which will reward you with success. You will have achieved personal independence from the gas pump and freedom from anxiety about shortages.

Good luck and happy motoring!

TYPES OF ALCOHOL FUEL

You probably have questions in your mind about using alcohol as a motor fuel. We'll look briefly into the history and types of alcohol fuels.

In the early days of the automotive industry several manufacturers produced motors designed to run on alcohol, instead of the relatively new and scare product called gasoline. The Ford Model T was originally designed to burn alcohol, gasoline or a combination of both. However, Ford and other manufacturers soon bowed to intensive pressure exerted by the emerging oil monopolies of the time and began designing motors with gasoline as the only intended fuel.

Several makers of farm tractors have continued production of alcohol engines. They are for export only, and you will probably never see a factory-engineered alochol tractor in this country at this time.

There are several types of alcohol. Each type is derived by a slightly different method or from a different raw material. We are interested in two types—methanol and ethanol.

Methanol may be familiar to you as the fuel used in Indy-type racecars. At one time, it was distilled from wood and commonly known as wood alcohol. Petroleum is the source of most methanol today. It is very toxic and poisoning can result from ingestion. Methanol is one possible source of fuel for your car, although it yields less energy than ethanol.

14

Ethanol is the best choice to run your car. It is formed by the fermentation of sugars from a number of types of organic materials. Wheat, corn, rice, beets and cane are examples of materials with a high sugar content. They give a high yield when fermented and distilled. All types of liquors contain ethyl alcohol also known as grain alcohol.

HOME DISTILLATION

The most economical way to obtain alcohol is to build a still and distill it yourself. You can legally distill alcohol fuel (but not alcohol to drink) by obtaining a permit from the Bureau of Alcohol, Tobacco & Firearms.

Homemade alcohol can be distilled for as low as 40¢ per gallon depending on the size of the still, the fuel used to power it and the cost of the grain. An increasing number of enterprising individuals, particularly farmers, are cutting their fuel bills by distilling their homegrown crops.

If distilling sounds attractive to you, there are several publications available on the topic. A good, basic primer on distillation is FORGET THE GAS PUMPS—MAKE YOUR OWN FUEL written by Jim Wortham and Barbara Whitener. It provides diagrams of small stills, step-by-step directions on how to make alcohol and the information necessary to obtain a permit. A model application is also included.

BUYING ALCOHOL

Industrial ethanol can be obtained from a wholesale chemical supply company. The price per gallon varies widely according to the quantity purchased. I purchased a 55-gallon drum of 200-proof ethanol for $124 or about $2.25 per gallon. Bulk quantities are more economical, but you must have a tank or other provisions for storage. Also, you must buy a large quantity to achieve the best price.

Check out all possible suppliers in your area, as there seems to be a wide variation in price from company to company. Comparison shopping may save you money.

Any alcohol you may purchase commercially has been denatured. Denaturing is a process required by the government to render alcohol unfit for consumption. Treat denatured alcohol as you would gasoline or any poisonous substance.

Figure 2-1—Here's how I "fill 'er up" with alcohol. If my tank had not been portable, a hose would have reached to the car.

PREPARING YOUR CAR FOR CONVERSION

Let's assume you have decided to convert "Old Betsy" to alcohol power. Maybe that old clunker in the yard— the one that barely runs, is undependable and isn't worth trading in. A second car not necessary for daily use is a dandy choice for initial conversion. You won't be without transportation during the workweek and working on an older car may ease those fears about how alcohol may adversely affect the engine of your later model car.

Just keep this one fact in mind; if it won't run on gasoline, you're defeated before you started. Alcohol will not magically cure a sick engine.

Take a realistic look at the condition of your engine. My test vehicle required replacement of a bad starter, old spark plugs and cracked spark plug wires. If you intend to keep the car for any length of time, the expense of major repairs will be justified.

When your car is running reasonably well on gasoline, it is ready to be converted.

MODIFICATION
OF THE FUEL LINE

A decision will have to be made at this point. Is the conversion going to be a permanent one; or do you want to use gasoline when readily available to reserve your car's new capability to burn alcohol for possible future shortages?

In a permanent conversion, the gasoline tank will become an alcohol tank. This involves running the tank completely dry and replacing the gasoline with alcohol. It is undesirable to mix alcohol with gasoline. They will not mix unless the alcohol contains almost no water.

I decided to install a dual-fuel system since the car would be run primarily on gasoline when returned to the owners.

The alcohol tank was constructed from a 5-gallon heavy-walled Nalgene plastic container. This container was obtained from a laboratory at no charge. Since it was small enough to be portable, I did not permanently mount it. The level of liquid could be seen at a glance since the plastic is translucent. A metal container would have served the purpose, but would not have been quite as handy.

A larger alcohol tank could be mounted on the trunk or roof of the car. The choice will depend on what you have to work with and your individual preference. By all means, pick a container that is sturdy, leakproof and clean.

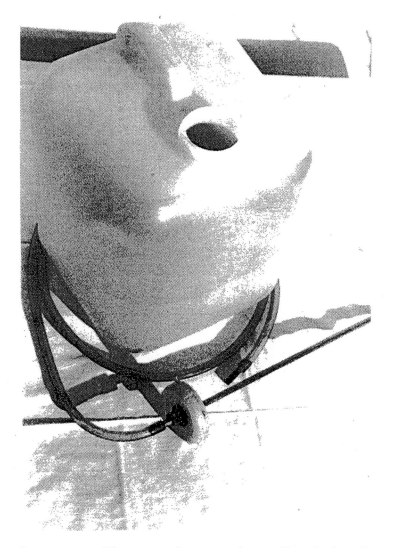

Figure 4-1—This is my homemade 5-gallon fuel tank and pickup tube. Everything shown was scrounged up at no cost whatsoever.

Install a pickup tube of the same outside diameter as your existing fuel lines. The tube should reach to approximately 1/16" from the bottom of the tank and protrude through the top approximately 1". Also install a vent tube to allow air to replace the liquid as it is consumed. Without the vent vacuum will collapse the tank. Make suitable holes in the tank for the vent and fuel lines and epoxy or braze them in place. I brazed my pickup tube to a pipe plug since the filler cap was already tapped, to receive a threaded plug. The exact method will depend on what material you use.

Next, mount the tank in the selected location on your car. A large tank should be fastened or strapped securely to prevent shifting or overturning. My small tank was wedged in between the back seats and could not move at all. I don't recommend that you mount your tank in an enclosed space, however.

With the tank securely mounted, some provision must be made for connecting into the existing fuel line. For this purpose, I used a "T" valve. This device can be purchased at any recreational vehicle, supply house. The price will range from about $5 for a simple, manual valve to $40 or more for an electric valve remotely controlled from the dash.

My "T" valve was donated by a neighbor who had removed it from his truck. He had replaced it with a more costly, electric valve. This particular model had three inlets. Only two inlets were required, and I plugged the third one.

I mounted the valve on the inner fender well, as close to the fuel pump as was practical. It could also be mounted under the floorboard so the handle could be operated from within the car. This arrangement would eliminate the need to raise the hood in order to change from one fuel to another.

I used standard black fuel hose for all necessary rerouting of fuel lines. Standard barb fittings were used to connect the hose to the "T" valve. Teflon tape was used on all threaded connections to prevent leaks.

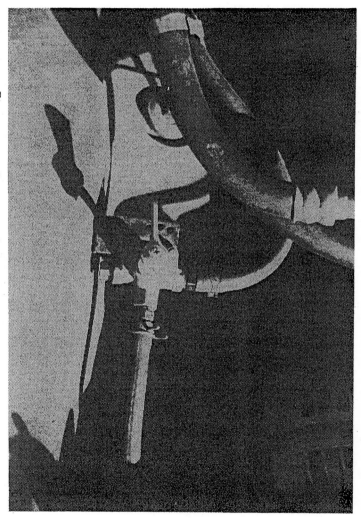

Figure 4-2—I mounted my "T" valve on the inner fender. The line on top is the alcohol inlet. The right line is the gasoline inlet.

Here's how to install your "T" valve and change the fuel lines:

1. Mount the "T" valve in the location you prefer.
2. Install barb fittings, if required.
3. Connect the alcohol tank to one inlet of the "T" valve using black fuel hose of the correct size.
4. Locate the inlet side of the fuel pump and remove the short piece of hose between the pump and the metal, fuel line from the gasoline tank.
5. Connect the metal fuel line to the remaining inlet on the "T" valve with black, fuel hose.
6. Connect the outlet of the "T" valve to the inlet side of the fuel pump with a third length of black fuel hose.
7. Make certain the lines you have added are secure, leakproof and do not interfere with any moving parts.

You now have a dual-fuel system, interchangeable at the flick of a valve.

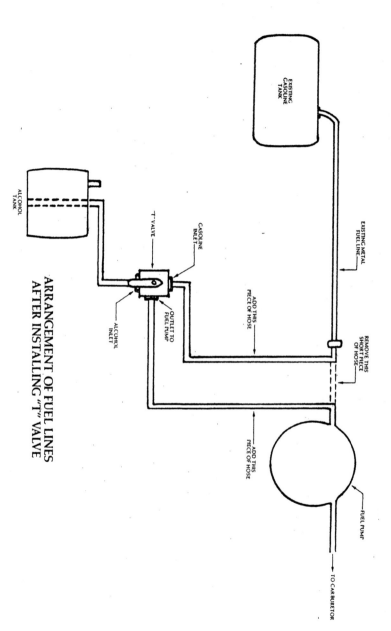

**ARRANGEMENT OF FUEL LINES
AFTER INSTALLING "T" VALVE**

EXISTING
GASOLINE
TANK

ALCOHOL
TANK

EXISTING METAL
FUEL LINE

"T" VALVE

GASOLINE
INLET

ADD THIS
PIECE OF HOSE

ALCOHOL
INLET

OUTLET TO
FUEL PUMP

REMOVE THIS
SHORT PIECE
OF HOSE

ADD THIS
PIECE OF HOSE

FUEL PUMP

TO CARBURETOR

Figure 4-3—The above diagram describes how I arranged
the fuel lines on the Dart.

THE CARBURETOR

TYPES OF CARBURETORS

The carburetor is to your car's engine what your heart is to your body. Its job is to feed fuel and air to the cylinders in the correct amounts.

Figure 5 - 1 is a simple diagram of a carburetor. It has one bore or barrel for air and fuel to pass into the intake manifold and one main metering jet to measure out fuel. This type works well on engines with a single bank of cylinders, such as a straight six-cylinder engine.

A "two-barrel" carburetor has two bored and two main metering jets. Each barrel is responsible for the cylinders on one side of a V-type engine such as a V-8. A better distribution of fuel is obtained with this arrangement.

Large displacement engines require more fuel, under some conditions, than a two-barrel carburetor can provide. If the "two-barrel" were large enough to meet peak demand, it would waste fuel under a light load. The "four-barrel" carburetor has two small primary barrels and metering jets, plus two additional secondary barrels and jets. The larger secondaries remain closed until the engine needs the extra boost they provide.

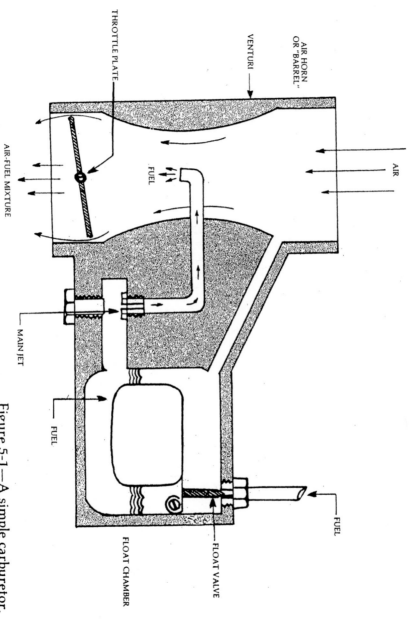

THROTTLE PLATE

VENTURI

AIR HORN OR "BARREL"

AIR-FUEL MIXTURE

FUEL

AIR

MAIN JET

FUEL

FUEL

FLOAT CHAMBER

FLOAT VALVE

Figure 5-1—A simple carburetor.

HOW THEY WORK

Whichever type your car may have, the operation is similar.

Fuel is pumped through the line and enters the carburetor's float bowl. A needle valve admits fuel to the bowl and closes off the flow when the bowl is filled. The valve is opened and closed by a float which rises and falls with the fuel level in the bowl. The fuel level is very critical and must be kept within close limits by the float and needle valve.

Air enters the carburetor through the air horn. It is drawn down past the venturi, where the passage narrows. This restriction causes an increase in air velocity and results in the creation of a vacuum below the venturi.

This vacuum sucks fuel from the bowl through the metering jets. The jets are carefully sized to give the correct ratio of fuel to a given volume of air. Metering rods may be used to partly close the main jets at idle.

The metered fuel passes into the venturi and is misted into the barrel. Here it is mixed with air and distributed to the cylinders by way of the intake manifold.

A smaller idle passage is equipped with an adjustable jet. The smoothest idle can be obtained by screwing the stem of this jet in or out.

A third fuel passage leads from the accelerator pump, located next to the float bowl. This passage provides an extra shot of fuel only when the accelerator pedal is pressed down. The extra fuel balances the excess air

which is sucked in as the throttle plate is suddenly opened during acceleration. The accelerator pump is inactive at idle or when the accelerator pedal is held at a steady position.

The choke valve is located in the air horn above the venturi. Its purpose is to restrict the amount of air entering the carburetor while the engine is cold. Fuel will not vaporize well in a cold engine and less air flow is desirable. An automatic choke will open and admit more air as the engine heats up. It is necessary to open a manual choke by hand.

A gasoline engine runs best with an air/fuel ratio of about 14/1. The jets in the carburetor are sized to provide this ratio.

Alcohol burns most efficiently with a ratio of around 9/1. The next chapter will outline the procedure for removing, modifying and replacing the carburetor to attain this ratio.

ALTERING THE
CARBURETOR

You may want to obtain a second carburetor to convert and keep your present one for gasoline use. I obtained a used one for $10 from a salvage yard. Rebuilt carburetors are available from your auto parts dealer. In either case, make sure the one you purchase is compatible with the engine you plan to use it on. Your parts dealer will be helpful in determining the most efficient.

Next, get a carburetor overhaul kit. Be sure the kit is for your particular model carburetor. The model will be cast into the body or on a metal tag attached to it. The kit contains new gaskets, check valves, an accelerator pump plunger and the necessary parts to make an old carburetor perform like new. Best of all, it has detailed instructions and an exploded view of all the parts. Each part is numbered and named for reference. This will be a real aid for reassembly.

REMOVE AND DISASSEMBLE THE CARBURETOR

Take a good look at your engine before removing the carburetor. Notice how each linkage or hose is attached. Remember—everything must go back on correctly, so be alert. Unbolt the carburetor and put any small fasteners away where they won't get lost.

Figure 6-1—This is the carburetor with the air horn and float chamber cover removed. The jets are lying just below the carburetor body.

Figure 6-2—Here's the carburetor disassembled a step further. The jets, metering rods and float have been removed. Be sure you know where each part goes before you remove it from your carburetor.

ENLARGING THE METERING JETS

Try to find a clean, uncluttered table or counter to work on. Use the exploded view as a guide and disassemble the carburetor far enough to remove the main, metering jets. The jets are more accessible in some models than others.

Carefully determine the hole size of the metering jets. This can be done by actually measuring the jet or looking in a shop manual. If you have no way of determining the hole size, take the jets to an automotive machine shop. They have the tools to measure and enlarge the jets.

Determine the proper jet size by the following example. This is the calculation I used for the Dodge.

Original Jet Diameter:	$d =$ .062
Increase Jet Diameter	.062
By 40% For Alcohol:	x 1.4
	248
	062
Correct Jet Diameter	
For Alcohol:	.0868

NOTE: The 40% figure used here will assure a rich enough mixture to prevent burning the valves due to an over-lean mixture.

After testing the car, you may find the jets are too large and the mixture is very rich. If this is the case, install new jets enlarged 35% or even smaller as your car requires. Just decrease the size in increments of 5% over stock size to be safe.

DECIMAL EQUIVALENTS AND TAP DRILL SIZES

FRACTION OR DRILL SIZE	DECIMAL EQUIVALENT	TAP SIZE	FRACTION OR DRILL SIZE	DECIMAL EQUIVALENT	TAP SIZE
NUMBER SIZE DRILLS 80	.0135			39	.0995	
79	.0145			38	.1015	5 40
1/64	.0156			37	.1040	5 44
78	.0160			36	.1065	6 32
77	.0180		7/64	.1094	
76	.0200			35	.1100	
75	.0210			34	.1110	
74	.0225			33	.1130	6 40
73	.0240			32	.1160	
72	.0250			31	.1200	
71	.0260		1/8	.1250	
70	.0280			30	.1285	
69	.0292			29	.1360	8 32,36
68	.0310			28	.1405	
1/32	.0312		9/64	.1406	
67	.0320			27	.1440	
66	.0330			26	.1470	
65	.0350			25	.1495	10-24
64	.0360			24	.1520	
63	.0370			23	.1540	
62	.0380		5/32	.1562	
61	.0390			22	.1570	
60	.0400			21	.1590	10 32
59	.0410			20	.1610	
58	.0420			19	.1660	
57	.0430			18	.1695	
56	.0465		11/64	.1719	
3/64	.0469	0 80		17	.1730	
55	.0520			16	.1770	12 24
54	.0550			15	.1800	
53	.0595	1 64,72		14	.1820	12 28
1/16	.0625			13	.1850	
52	.0635		3/16	.1875	
51	.0670			12	.1890	
50	.0700	2 56,64		11	.1910	
49	.0730			10	.1935	
48	.0760			9	.1960	
5/64	.0781			8	.1990	
47	.0785	3-48		7	.2010	1/4-20
46	.0810		13/64	.2031	
45	.0820	3 56		6	.2040	
44	.0860			5	.2055	
43	.0890	4 40		4	.2090	
42	.0935	4 48		3	.2130	1/4 28
3/32	.0938		7/32	.2188	
41	.0960			2	.2210	
40	.0980		LETTER SIZE DRILLS 1	.2280	
			A	.2340	

Figure 6-3—A standard machinist's drill chart.

Next I looked up .0868 in a standard drill size chart. This chart can be found in most auto repair manuals or machinists handbooks. The closest drill size is a #44 which is .0860.

Once you have determined the proper drill size, use it to carefully enlarge each jet. A drill press is best for this job, but a hand drill can be used if caution is exercised. **Do not** attempt to drill the jets without removing them from the carburetor. The shavings will fall into tiny passages and create problems.

You may be able to avoid drilling. Auto parts dealers stock jets in various sizes for many types of carburetors.
They are inexpensive and if you can find one in the proper size, it will save time and labor.

I opted for the drilling method.

Figure 6-4—This is a typical set up to drill the jets on a small drill press. The "V" block holds the jet parallel with the drill bit.

THE FLOAT—TWO POSSIBLE ALTERATIONS

Alcohol is heavier and more dense than gasoline. The float will ride higher in a given amount of alcohol than in the same amount of gasoline. Since fuel flow is cut off when the float rises to a specific point, the alcohol level will be lower than if the float bowl were filled with gasoline. Some method must be used to raise the fuel level to normal. Otherwise, the engine will cut out during turns due to a lack of sufficient fuel.

BENDING THE FLOAT LINKAGE

One way is to simply bend the linkage between the float and the inlet valve.

The exact amount to bend the linkage is subject to trial and error. Several settings may be required to find the correct float height.

ADDING WEIGHT TO THE FLOAT

A difficult but more accurate method is to add weight to the float. Carefully weigh the whole float assembly on a balance (a powder scale will do an excellent job). Weigh a piece of solder or epoxy glue equal to 10% of the total weight of the float assembly. Fasten this to the top of the float, distributing the weight as evenly as possible.

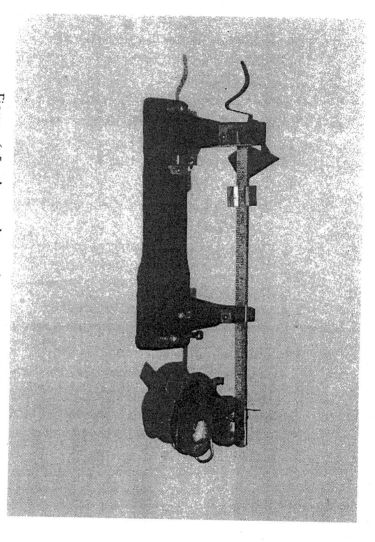

Figure 6-5—A powder scale is just the thing for weighing the float. My float weighed 253 grains.

I tried both methods. The bending method was tedious and involved disassembly of the carburetor several times before the correct setting was found. Also, some float bowls don't have enough clearance above the float to allow it to be bent enough to do any good.

The weight method worked just fine. It is preferable because you can set the float just as the instructions in the overhaul kit call for.

Reassemble the carburetor. Follow the kit guidelines for reassembly and adjustment.

Got any leftover parts? Hopefully, your answer is no! Now replace the carburetor on the engine and connect everything back up.

ALTERING THE IGNITION SYSTEM

Once the fuel system mixes the proper ratio of air to fuel and delivers the mixture to the individual cylinders, its function is complete.

The actual firing and production of useful power is controlled by the ignition system. Its task is to ignite the fuel/air mixture at a precise instant in relation to the movement of the piston. If the mixture fires too early, pre-ignition pinging will occur and may damage the engine. If the mixture ignites too late, power will be lost and fuel wasted.

A gasoline engine is timed to fire just as the piston reaches the top of its stroke or just an instant before. Gasoline ignites and burns so rapidly that the result is more like an explosion than a burning.

Alcohol burns at a much more even rate. It is harder to ignite and requires more time to completely vaporize and burn. To allow for this extra time, it is necessary to "advance" (cause to fire earlier) the timing of the engine. This is done by loosening the distributor and turning it in the opposite direction from the direction the rotor in it turns.

If you aren't sure which way the rotor in your engine turns, take the distributor cap off and have an assistant crank the engine. Whichever way the rotor turns, you must turn the distributor body the opposite way to advance the timing..

A timing light will make this job a snap, if you have access to one. The engine can be timed by ear, if you develop the knack for it.

Here's how to do it—

1. Loosen the bolt at the base of the distributor until the whole unit can be turnèd.

2. Initially advance the timing by turning the distributor about 1/16 of a turn. Grasp the cap and turn the whole thing. Snug the hold-down bolt so the distributor won't turn itself (but you can still move it with a little effort.)

3. Start the engine and run any gasoline remaining out of the fuel line. Make sure it's running on alcohol. Now advance still more until you get a faster idle, but still a smooth idle.

4. Check the timing by driving. If you can hear "pinging" in the engine during acceleration, the timing must be "retarded". The correct setting will be just below the point where pinging disappears.

I initially timed the Dart by ear and made several changes in timing before I was completely satisfied. After several days of driving, I checked the timing with a light. The original setting had been 0° TDC (top dead center). After converting to alcohol, the optimum setting had become 24° BTDC (before top dead center).

Your engine may not require **exactly** 24° of advancement. This is a ballpark figure. The exact amount will depend on your particular engine.

SPARK PLUGS CAN MAKE A DIFFERENCE

Let's discuss spark plugs briefly. As you may already know, each type spark plug is made in three or four heat ranges. The only difference between two plugs of the same **type** but different **heat range** is the rate at which they dissipate heat. The plug with the higher number will hold heat longer and is referred to as a **hotter** plug.

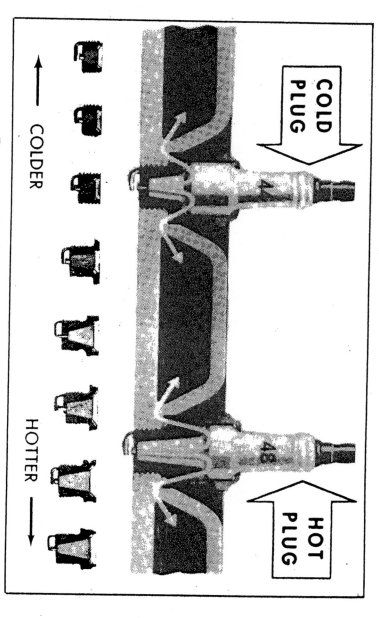

Figure 7-1—Spark plug heat range system. The lower number denotes the "colder" type plug.

Alcohol fuels, as I mentioned before, don't ignite as easily or burn as quickly as gasoline. To increase the temperature in the firing chamber, it may be desirable to use a plug that is one or even two ranges better than your engine has been using.

The Dart originally had Champion N-14-y spark plugs. After conversion, the engine was prone to miss due to poor vaporization of the fuel. More heat was needed to turn the alcohol into a combustible gas.

After installing a hotter set of N-16-y plugs (I had to go up two heat ranges), I could tell a definite difference. The increased heat was getting more fuel to completely vaporize and cutting the time it took for the engine to warm up.

Be careful not to go to extremes and install a plug which is too hot. The telltale sign will be a white or blistered-appearing electrode. Check your plugs often for the first few miles until you are certain they are correct.

FIRING UP
THE ENGINE

At this point, you should be ready to start up. The carburetor is modified and back in position. Timing should be advanced to the first tentative setting. Any necessary fuel line changes have been completed.

Prime the carburetor with a few drops of gasoline, fill the alcohol tank and crank the engine. You may have to "feather" the gas pedal to keep it running once the alcohol begins to reach the engine. (The idle mixture is still set for gasoline at this point.)

Let the engine warm up before you set the idle. Then slowly back the idle adjusting screws out 1/4th turn at a time until a smooth idle is obtained. If there are two idle screws, move both the same amount each time.

Get in the car and test drive it. Notice any symptoms which seem irregular. The chart I have provided lists some symptoms I have encountered and the possible causes and alterations.

SYMPTOM	PROBABLE CAUSE	ALTERATION
Engine cuts out under acceleration	1. Low float level 2. Accelerator pump not at a full stroke.	1. Raise float or add more weight 2. Adjust accelerator pump for a longer stroke.
Engine dies when turning a corner	Low float level	Raise float or add more weight
Pinging under hard acceleration	Timing set too advanced	Retard timing until symptoms stop.
Engine dies at idle	1. Carburetor improperly adjusted· 2. Spark plugs are incorrect heat range if wet—too cold, if white—too hot.	1. Adjust as advised in alcohol kit and this text. 2. Change plugs to proper heat range
Engine will not start when cold	Poor fuel vaporization qualities	1. Read the next chapter for solution. 2. Proof of alcohol may be too low. Use 160 proof or better

COLD WEATHER STARTING PROBLEMS

HOW TO CURE THEM

Alcohol does not readily vaporize below about 70°F. Unless you live in a tropical climate, some provision will have to be made for starting the engine in cold weather.

STARTING THE ENGINE ON GASOLINE

If you installed the "T" valve type fuel system, you can start the car on gasoline. I used this system in the Dart and had no significant starting problems. Each time I parked the car for an extended period, I would switch the "T" valve to gasoline just a minute or so before shutting the engine off. The gasoline burned very rich, but would burn. In the morning, I would start the car and switch back to alcohol in a minute or so. A gallon of gas would last a week or so with this type of starting method. (Mind you I said a gallon of gas—not a tank.)

In practical use, the smaller tank could be one gallon or less in size and alcohol could go directly into the regular tank. Alcohol will clean any old gunk out of the tank. You may have to change the fuel filter several times until the tank has been cleaned by the alcohol.

USING GASOLINE AS A PRIMER

Gasoline is also used in another method. This method involves installing a surplus windshield washer pump and reservoir. The pump has a hose leading to a tube or spray nozzle brazed into the air cleaner lid. The reservoir is then filled with gasoline.

A few squirts of gasoline will fire up the cold engine. Once the engine has generated a little heat, it will run well on alcohol.

PREHEATING THE ALCOHOL

Several solutions involving preheating of the fuel have been found. Most involve applying an external source of heat to the carburetor or the fuel line. Any method of raising the fuel temperature above 70°F will be most helpful.

Your ingenuity can get a workout in this area. Don't be afraid to experiment with any ideas you may have which will hold in engine heat and ease starting.

MANUAL CHOKE

Another aid which I found helpful is the good old-fashioned hand choke. These are available at auto parts stores and will replace your automatic choke. The hand choke will allow "pampering" of a "finicky" cold engine. It will also allow you to open the choke wide to take advantage of the window-washer method of priming.

VITAL
STATISTICS

MATERIALS AND COST OF DART CONVERSION

Used carburetor	$10.00
Manual choke kit	2.49
Eight (8) feet of fuel hose	4.80
"T" valve (free)	
Five (5) gallon tank (free)	
Three (3) barb fittings for fuel lines	1.26
Eight (8) N-16-y Champion spark plugs	8.80
One (1) roll teflon tape (for fittings)	.69
Overhaul kit for carburetor	7.00
Total parts cost	$35.04
Ky. 5% Sales Tax	1.75
Labor-Supplied by writer	
Total	$36.79

SPECIFICATIONS OF DART CONVERSION

Automobile tested   1969 Dodge Dart
Mileage at time of conversion   79950 miles
Engine type   318 C.I.D. V-8
Carburetor type   Carter Model BBD 2BBL.
Fuels tested   Ethyl alcohol 150/200 proof
Best proof
  (Determined by test)   180 proof
Miles per gallon-180 proof   11 miles
Miles per gallon-Gasoline   13.5 miles
Price per gallon-200 proof $2.25
Price per gallon-180 proof (diluted) $2.07
Percentage of fuel cost for imported oil   0%
Writer's Satisfaction   100%

MY EXPERIENCE IS IN YOUR BALLPARK

This guide is based on my own experiences with an older model car. I performed all modifications as described in this text. Your results may be much improved over mine and you might hit upon a better idea for easier starting and increasing mileage.

Many other gasoline-powered implements can be converted to alcohol. Some of these include:

1. Motorcycles

2. Boats

3. Lawnmowers

4. Emergency Generators

5. Farm Implements

I plan to continue to experiment, and hopefully, convert these also as the opportunity comes about.

Drop me a note and let me know how your experiences with alcohol fuel turn out. I will try to answer any questions you may have and will be interested to hear of your successes in conversion to alcohol fuel.